THE BIG IDEA:
CURIE & RADIOACTIVITY

Paul Strathern was born in London. He has lectured in philosophy and mathematics at Kingston University and is the author of the highly successful series *The Philosophers in 90 minutes*. He has written five novels (*A Season in Abyssinia* won a Somerset Maugham Award) and has also been a travel writer. Paul Strathern previously worked as a freelance journalist, writing for the *Observer, Daily Telegraph* and *Irish Times*. He has one daughter and lives in London.

President Kennedy School Library

D1427091

PET

Curie & Radioactivity

PAUL STRATHERN

ARROW

Published in the United Kingdom in 1998 by
Arrow Books

1 3 5 7 9 10 8 6 4 2

Copyright © Paul Strathern, 1998

All rights reserved

The right of Paul Strathern to be identified as the author of
this work has been asserted by him in accordance with the
Copyright, Designs and Patents Act, 1988

This book is sold subject to the condition that it shall not, by
way of trade or otherwise, be lent, resold, hired out, or otherwise circulated without the publisher's prior consent in any
form of binding or cover other than that in which it is
published and without a similar condition including this
condition being imposed on the subsequent purchaser

First published in the United Kingdom
in 1998 by Arrow Books

Arrow Books Limited
Random House UK Limited
20 Vauxhall Bridge Road, London SW1V 2SA

Random House Australia (Pty) Limited
20 Alfred Street, Milsons Point, Sydney,
New South Wales 2061, Australia

Random House New Zealand Limited
18 Poland Road, Glenfield
Auckland 10, New Zealand

Random House South Africa (Pty) Limited
Endulini, 5a Jubilee Road, Parktown 2193, South Africa

Random House UK Limited Reg. No. 954009

A CIP catalogue record for this book
is available from the British Library

Papers used by Random House UK Limited are natural,
recyclable products made from wood grown in sustainable
forests. The manufacturing processes conform to the en-
vironmental regulations of the country of origin

Typeset in Bembo by SX Composing DTP, Rayleigh, Essex
Printed and bound in the United Kingdom by
Cox & Wyman Ltd, Reading, Berks.

ISBN 0 09 923842 X

CONTENTS

INTRODUCTION

Marie Curie was the 20th century's most exceptional woman. Her discoveries won her two Nobel prizes for science (a feat not equalled for over half a century). Her consequent work furthering the cause of radium research resulted in major advances in nuclear physics and in the use of radium therapy treatment for cancer. Both her husband Pierre Curie and her daughter Irène Joliot-Curie also won the Nobel prize. Marie Curie finally died of leukaemia resulting from years of work in a primitive laboratory isolating radium. It all sounds a bit too good to be true.

Little wonder that the world was willing to accept the picture of a secular saint drawn by her daughter Eve, in her dutiful biography published four years after her mother's death. This book proved inspirational to many women in their fight for recognition – as women, as independent spirits, and as scientists. But it also portrayed one of the most perfectly boring women imaginable. Fortunately, the real Marie Curie was nothing of the kind. As we now know, she was a highly passionate woman – both in her work and her life. Desperately unlucky in love, she had sufficient strength not only to resist the temptations of money and fame, but also the execration of public scandal. (She was one of the first to suffer the full tabloid treatment.) To depict Marie Curie as a saint is to malign her. She was a single mother, who brought up two daughters, and made a major contribution to 20th century science.

LIFE & WORK

Marie Curie was born Maria Sklodowska in Warsaw on 7th November 1867, the youngest of five children. Her father was a schoolteacher, specializing in physics and maths. Her mother was headmistress of the best private girls' school in Warsaw, and the family lived in the apartment behind the school on Freta Street.

These were difficult times in Poland, which was under Russian rule. Following the widespread but unsuccessful uprising of 1863, over 100,000 Poles had left the country. Many had gone into exile in such places as Paris and North America, whilst others had been forcibly shipped to Siberia. After this, Russian rule had become increasingly oppressive: public hangings were still being conducted at the citadel in central Warsaw at the time of Maria's birth.

Around 1870 Maria's mother contracted TB. At the same time her father was demoted at school – largely because he was a Pole, but also because he was (correctly) suspected of sharing his nationalist principles with his pupils. Money was now short in the family, but worse was to come. In 1878, when Maria was ten, her mother died of TB and her father was dismissed. The family were forced to take in boarders, simply to make ends meet. Maria slept in the living room – doing her homework after the others had gone to bed, and rising early to lay the table for the boarders' breakfast.

Photos of the period show Maria as a plain intense girl. She had the chubby cheeks of her mother, restrained fluffy curls, and thick, slightly pursed lips. But her appearance was almost the only ordinary thing about her. At school, where she was forced to study in a foreign language (Russian), she demonstrated exceptional ability. She graduated a year early, at 15, carrying off the gold medal. And that was it. There was no further education for girls in Poland.

Maria was looking a little wan after all her exertions, so she was sent to stay with her uncles.

They were remnant members of the landed gentry, with small estates way out in the middle of nowhere close to the Ukrainian border. Here Maria found herself in 'oases of civilization in a land of rustics'. For the first time in her life (and the last) she lived a happy, utterly carefree life. Aunt Maria was a liberated woman, and expected her daughters to be strong and independent. Young Maria and her cousins visited the neighbouring houses of the surprisingly cultured local gentry. Here they played music and read French and Polish literature to one another – a heady brew containing the likes of Chopin and Victor Hugo, as well as the great Polish Romantic poet Mickiewicz and Slowacki the Polish Byron (both of whom had recently died in exile). On high days and holidays Maria and her cousins would attend country gatherings in local costume, often dancing long into the early hours. This continued for almost a year.

When Maria finally returned to Warsaw, she found that her father had lost what little money he had through unsound investments. The family were living in near poverty and Maria took work as a teacher, contributing her wages to the depleted family coffers. But she also made contact with the illegal Polish 'free university', which was a 'wandering' institution (ie, it moved from place to place to evade detection by the Russian authorities). As was the practice, she gave as well as received. In return for books to read, and occasional lectures, she read to women workers, instilling in them their Polish heritage. At the free university, socialism, science and scepticism were the order of the day, and Maria soon lost any remnant of religious belief. She began reading widely, in a variety of languages: Karl Marx in German, Dostoevsky in Russian, and poetry in French, German, Russian and Polish. She even tried writing her own poetry, and worked for the underground magazine *Prawda* (meaning 'truth'; not to be confused with the later Russian version, which peddled the opposite).

Fortunately *Prawda* was devoted to the new religion of science, and Maria soon saw the light. The cryptic algebra and banal formulae of poetry gradually gave way to the soaring poetry of pure mathematics and the romanticism of scientific discovery. Maria had found her subject. But what was she going to do about it? Where could she study it to some purpose?

Maria entered into a pact with her older sister Bronia, who wanted to study medicine. She would work in Poland to finance Bronia's studies in Paris, and then in return Bronia would help her to study science in Paris.

Bronia set off for Paris, and Maria took a post as a governess in the home of a well-off estate manager in the country, 60 miles south of Warsaw. Maria's job was to educate the two daughters of the family, one of whom was her own age. But this was to be no cultural oasis in a rural idyll. As the paltry joys of the beetroot harvest festival gave way to the frozen muddy wasteland of winter, Maria became appalled at the poverty and ignorance of the local peasants. Mindful of her free university training she set up a class to teach the peasant children their XYZ in the Polish language. As if this wasn't enough, she also continued with her own self-education. 'At nine in the evening,' she wrote to her sister, 'I take my books and go to work . . . I have even acquired the habit of getting up at six so that I work more.' She reports that she is reading no less than three books at a time: Daniel's *Physics* 'of which I have finished the first volume', Spencer's *Sociology* in French, and Paul Bers' *Lessons on Anatomy and Physiology* in Russian. 'When I feel myself quite unable to read with profit, I work on problems of algebra or trigonometry, which allow no lapses of attention and get me back into the right road.'

All this may sound a bit too much to believe, but there's no doubting that Maria studied hard during the long snow-bound winter nights. From her early days sleeping in the living room she had been used to fighting for time to study. And now at last she had a goal: Paris. If she buried herself in work, the three years of governess-drudgery would pass even quicker, and she would arrive in France all that better prepared.

But even the dullest and most determined little swot is prey to fits of normality. The ice fields thawed, giving way to waving fields of purple-green beetroot bloom, heralding the long hot days of summer. The Zorawski's eldest son arrived home for his holidays. Kazimierz was a mathematics student at Warsaw University, a year older than Maria. According to her letters, none of the other young men in the area were 'even a bit intelligent'. So lightning struck, as it will. Maria and Kazimierz fell in love.

By the time Kazimierz came home for his Christmas holidays they were talking of marriage. Then the parents got wind of what was going on between their dear Kaziu and that earnest little governess, who was not only plain but penniless. Marriage to such a declassé creature was quite out of the question for the Zorawski son and heir. Tamely (and possibly with a little relief), the 19-year-old Kazimierz submitted to his parents' wishes. The romance was over: Maria was devastated. But she was strong enough, and independent enough, to keep her feelings largely to herself.

Her suffering can only be imagined, as Maria gritted her teeth and continued to work out the years of her contract. Why didn't she just leave? Each holiday Kazimierz returned from Warsaw, and against all the odds she still went on hoping. Year in, year out. When Bronia finally wrote from France with the news that she was planning to marry a fellow medical student, which meant Maria would eventually be able to come to Paris and stay with her, she even hesitated. Despite her previous, almost obsessive determination, she would have been willing to give it all up for Kazimierz.

But there was bitterness too. When Maria heard that her other sister Helena had been rejected under similar circumstances, she at last found a legitimate vent for her anger. As she gradually gave way to her feelings, in an increasingly emotional letter (ostensibly expressing outrage at her sister's plight), we catch a glimpse of all that she has kept to herself: 'I can imagine how Hela's pride must have been humiliated . . . If they're not interested in marrying poor girls, they can go to hell . . . But why do they insist on upsetting such an innocent creature?' She ends with a curious, but revealing remark: 'But I, even I, keep a sort of hope that I shall not disappear completely into nothingness.'

Maria was aware of the clear outlines of her character – and how she appeared to others. Her dedication had required self-denial, her suffering had required self-suppression – but she was no nonentity. Maria Sklodowska was now more determined than ever to make something of her life. Her years as a governess had hardened her. She did her best to disguise this: 'I often hide my deep lack of gaiety under laughter.' But when she returned to her family in Warsaw, it was evident to them that something in her had changed. And this was more than just growing up, though she was now 22.

Maria spent another couple of years in Warsaw, working as a governess and saving every last *grosz*. Then in 1891 she at last set off for Paris. She was now 24 – an age when some of her great contemporaries were to be on the brink of major discoveries – and she hadn't even started her degree. (At the age of 25 Einstein was to discover relativity, Marconi would be sending radio signals across the English Channel, and Rutherford would be launched into nuclear physics.)

Maria took the train from Warsaw to Paris. She travelled fourth class, perching on a canvas camping stool by her luggage for the three-day journey. (The intellectual mecca of Paris proved a powerful attraction to penniless young travellers possessed of exceptional willpower and talent during these years. The French poet Rimbaud *walked* from Vienna – as did the

Romanian sculptor Brancusi from Bucharest. Such was the competition if you wished to succeed in the City of Light.)

Maria enrolled in the Faculté de Sciences at the Sorbonne (the University of Paris). There were 1,800 students, just 23 of whom were women and less than a third of these were French. In Paris the word for female student *étudiante* had much the same nudge-and-a-wink connotations as the word model today. No respectable father would subject his daughter to such humiliation, especially if this only made matters worse by educating her. Most Frenchmen agreed with the contemporary writer Octave Mirabeau: 'Woman is not a brain, she is a sex.' In Poland Maria had been able to develop her independence – not purely in the intellectual sphere, either. In Paris it was back to Neanderthal Man: any woman out on the streets at night was automatically a prostitute. The light which first illuminated the City of Light in 1891 was limited to the electrical sphere.

Maria began as she meant to go on. Politely but firmly refusing her sister's offer of accommodation, she went to live alone in a *chambre de bonne*, the traditional squalid garret room where geniuses starved in the Latin Quarter around the Sorbonne. After lectures, experiments in the lab and reading up in the library, she would climb the six floors to her room under the sloping roof. After dining off a *ficelle* (literally a 'string': the thinnest French loaf) and a slab of chocolate, she would work long into the night. Now at last she was completely free to follow her studies, and no one could stop her. In middle age she would recall this time as 'one of the best memories of my life'. This was the 'period of solitary years exclusively devoted to study . . . for which I had waited so long'. She even wrote a poem about it:

'Yet she has joy in what she knows
For in her lonely cell she finds
Rich air in which the spirit grows,
Inspired by the keenest minds.'

Paris had a thriving community of Polish exiles: virtually a cultural and political elite-in-waiting. Its calibre and ambiance are perhaps best epitomized by its most brilliant young member:

Paderewski – who was to become a world renowned concert pianist, Prime Minister of Poland, and a lover of Greta Garbo (though only consecutively). Maria avoided such frivolities: the stars of her firmament were French and scientists. Despite her love of her homeland, she identified with her adopted country to such an extent that she even 'frenchified' her name to Marie. France was her opportunity: she would take all it had to offer.

These were vintage years for science at the Sorbonne. Education and science were the religion of the Third Republic, and a new Sorbonne was under construction with vast amphitheatre lecture halls and modern well-equipped laboratories. The medieval stronghold of Scholasticism throughout Europe now banished theology to the fringes. Literature too was downgraded: this was just a pastime for informed men of culture. (Science had not always been in such favour in France. At the end of the previous century, during the Revolution, the great Lavoisier 'the Newton of chemistry' had been despatched to the guillotine with the words: 'France has no need of scientists.')

Marie's heroes were the giants of the Sorbonne lecture halls. But as she wrote: 'The influence of the professors on the students is due to their own love of science and to their personal qualities much more than to their authority. One of them would say to his students: "Don't trust what people teach you, and above all what *I* teach you."' Science was moving fast, and many of Marie Curie's lecturers were in the forefront of modern research.

Her lecturer in biological chemistry was Emile Duclaux, one of the earliest protagonists of Pasteur and his theory that diseases were spread by microbes. Duclaux's lectures were laying the foundations of a new field: microbiology. Her physics professor was Gabriel Lippmann, who was in the process of inventing colour photography. But by far the finest mind she came into contact with was Henri Poincaré, the greatest mathematician of the period. Each year it was his custom to deliver original lectures on a new branch of applied mathematics. His 1893 lectures on probability theory were far ahead of their time. Poincaré was already anticipating concepts which later became integral to statistical mechanics, especially with regard to 'chaos'. (This is where the mathematics describing a dynamical system becomes so complex that the elements within it cannot be computed or even defined, thus remaining random and unpredictable.) Although Marie's inclination was undoubtedly towards science, her mathematical abilities were of an almost similar high order. In her final *licence* exams she came top in physical science, and second in mathematics.

But Marie's student life was not quite so exclusively solitary as she would have us believe in her memoirs. In 1893, the year she took her *licence*, she became attracted to one of her French fellow students. His name was Lamotte, and she seems to have been drawn by his similarly intense attitude towards science. Marie was only interested in 'serious conversations concerning scientific problems'. However, we know from surviving letters that she did find time to confide her aspirations to Lamotte. Curiously, her ambitions were limited to her class work. At this stage all she dreamt of doing was returning to Poland, living with her father, and becoming a teacher. Fortunately her professors got wind of this colossal waste.

Marie was suffering from post-exam let-down, and a bit blue over the way Lamotte had left her for his home in the provinces. His final letter had ended in most ungallant and un-Gallic fashion: 'Always remember you have a friend. Adieu! M. Lamotte.' (It's unclear whether this 'M' is simply an initial – standing, say, for Michel – or is the usual abbreviation for 'Monsieur'. Either way, hardly a rapturous farewell.) However, Marie's spirits soon rose when she received a note from Professor Lippmann. In this he invited her to work as a research assistant in his laboratory. Late in 1893 Marie began research into the magnetic properties of steel, a mundane but absorbing task on which to cut her teeth.

Early the following year, whilst visiting the home of a Polish physicist, she was introduced to a reserved 35-year-old man with a neat beard and hair cropped *en brosse* (like a brush: the slightly longer French crew cut). Marie recalled: 'We began a conversation which soon became friendly. It first concerned certain scientific matters.' Almost immediately they discovered 'a surprising kinship, no doubt attributable to a certain likeness in the moral atmosphere in which we were both raised.' They were both equally intense, they were both outsiders, and they were intellectual equals.

Pierre Curie was nine years older than Marie Sklodowska, and had already produced major work. Like Marie, Curie had been brought up in a scientific family, where progressive ideas and lack of religious belief were the norm. From an early age Pierre Curie had been a 'dreamer', given to bouts of introspective contemplation when he appeared completely unaware of the world around him. At school he was not a success, and was judged to have a 'slow mind'. It was decided that he should be educated at home. Even so, his mind continued to wander, he wrote sloppily, and was liable to make mistakes in the agreement of genders. (This process, which so baffles English speakers, quickly becomes second nature to any normal French child.) But when Pierre concentrated his thought exclusively on one topic, it soon became clear that he had exceptional intellectual qualities. In an effort to salvage his education he was encouraged to develop this trait. The ugly duckling miraculously developed into a swan. At the age of 16 he went to the Sorbonne.

After university Pierre did experimental work with his brother Jacques. Together they discovered that certain non-conducting crystals (such as quartz) developed an electric charge if distorted. When a quartz crystal was subjected to pressure its opposite faces developed opposite charges. This phenomenon they named the piezoelectric effect, after the Greek *piezo* meaning 'to press'. By reversing this process, the Curie brothers discovered that when a quartz crystal was subjected to an electric charge, its crystal structure became deformed. If the potential of the electric charge was rapidly changed, the faces of the crystal rapidly vibrated. This could be used to produce ultrasonic sound (ie, sound waves whose frequencies are too high for human hearing), and is used nowadays in a wide range of instruments from microphones to pressure gauges. The Curie brothers themselves made use of the effect to construct a highly sensitive electrometer capable of measuring minute electric charges.

At the age of 32 Pierre Curie had been appointed head of the laboratory at the Paris School of Industrial Physics and Chemistry. This was hardly a prestigious post, but Curie was more interested in following his own experimental bent than money or prestige. Pierre Curie was averse to distractions of any sort. He firmly believed that a wife could only be a hindrance to a scientist.

At the time Pierre met Marie he was completing his doctorate on the effect of heat on magnetic properties. He had discovered that above a certain critical temperature any ferromagnetic substance (such as iron or nickel) will lose its ferromagnetic qualities. (This temperature is still known as the Curie Point.) Marie too was doing research in this field – which leads to the inescapable conclusion that the pair were drawn together by magnetism. They quickly became friends.

When Pierre called to see Marie in her garret, he was immediately impressed by her simple independent lifestyle, which dispensed with such formalities as a chaperone. Here indeed was a progressive woman of science. But this was to be no head-over-heels romance. Both were mindful of their precious independence, which led to hesitations on both sides. But eventually Pierre decided to grasp the nettle. He wrote to Marie asking 'if you would like to rent an apartment with me on Rue Mouffetard with windows overlooking a garden. This apartment is divided into two independent parts.' Neither believed in the conventional life: they were above such things. But this was an intellectual rather than an emotional stance. Both had used it to enable them to devote their lives to science rather than as a vehicle for social outrage, which they saw as a waste of time: 'It is excusable to be generous with everything except time.' In their exchange of letters Pierre confessed: 'I'm far away these days from the principles I lived by ten years ago.' He no longer always wore a blue shirt 'like the workers'. (Though whether he still stuck to the principle that a partner was a mere hindrance to a scientist is not mentioned.) Marie for her part still felt the tug of her native land. She had been to Poland for holidays, but this was not enough: it was time for her to return.

If there was to be any future for Pierre in the life of Marie, and vice versa, they were going to have to commit to each other. This much gradually became clear to both of them. And so Marie and Pierre were married in a (strictly civil, plain clothes) ceremony at a Town Hall in the suburbs. There were no conventional wedding gifts. Instead of chair covers, a useful pudding steamer, and a cuckoo clock, the couple bought a pair of brand-new modern *bicyclettes* – and set off to cycle round Brittany for their honeymoon. In the course of this they discovered a deep love for the countryside and a deep love for each other, both of which would last them their lifetime together.

On their return to Paris, the Curies settled into a small three-room flat on the Rue de la Glacière on the Left Bank. Pierre supported them both on his small salary. Meanwhile Marie studied for the *agrégation*, the French higher teacher's certificate, took a number of additional courses in theoretical physics, and even managed to continue with her research into magnetism. According to the myth carefully cultivated in her letters home (and perpetrated in her memoirs and the hagiography written by her daughter): 'We see nobody . . . and give ourselves no diversions.' However, they did manage to cycle out into the countryside at weekends, and they appear to have availed themselves of the consolations of life in what was then the world's most sophisticated city. Admittedly, the Curies were hardly doyens of *fin de siècle* Parisian society: the world of Degas, absinthe and *les grandes horizontales* (as the glamorous society courtesans were known). But the newly-weds seem to have enjoyed regular nights out in the Latin Quarter. They sat in the beamed dark in the new *cinématographe* watching men in top hats and women in long dresses hopping along the boulevards. They even went to the theatre: no freethinking conversation was complete without the obligatory reference to Ibsen or Strindberg. Yet at least in their attire they stuck to their ascetic principles. In those days everyone dressed up to go to the theatre – but not the Curies. Friends spoke of being 'amazed' on encountering the two scientists in their fashionless attire. (How much this was simply due to Parisian standards of taste, and how much to the Curies' complete lack of it, is difficult to tell.) Despite such occasional evenings of wild frivolity, Marie passed top in the *agrégation* in physics and second in maths. Then she became pregnant, and in September 1897 gave birth to her first daughter, Irène.

At home Marie and Pierre had become very close: they shared and dicussed everything that interested them. In other words, science. Pierre's research, Marie's theoretical physics course, experimental difficulties, scientific problems – all received the same intense concentration. Right from the start, their minds had a deep rapport. Each felt that the other could understand their problem as no other. Even after the baby was born, entire evenings were often spent analyzing the latest developments in science.

During the first years of the Curies' marriage science began to change beyond recognition: the physics of the 20th century was being born. The earliest signs were deceptively mundane. In the autumn of 1895 the German experimental physicist Wilhelm Röntgen (sometimes spelt Roentgen) was going over some previous experiments observing the phenomena of luminescence. He began passing an electric current through a partially evacuated glass tube (a cathode ray tube, similar to what nowadays provides the screen of a TV set).

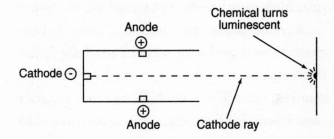

In his darkened laboratory at the University of Würzburg, Röntgen began investigating the luminescence which cathode rays induced in a number of chemicals. To assist his observations of this faint luminescence he placed the cathode ray tube in a blackened cardboard box. When he turned on the current a glimmer of luminescence caught his eye on the far side of the darkened room. It turned out to be a sheet of paper coated with barium platinocyanide (one of the luminescent chemicals which he had been testing). But how could this become luminescent when the cathode ray tube was placed in a box? The cathode rays should have been blocked by the blackened cardboard. He turned off the cathode ray tube, and the luminescence subsided. It was definitely caused by something to do with the cathode rays.

Acting on a hunch, Röntgen took the coated piece of paper into the next room, where he then pulled down the blinds and shut the door. When the cathode ray tube was turned on the paper glowed again. Some kind of unknown radiation was emanating from the cathode ray tube. This was not only invisible, but it was able to penetrate cardboard and other materials. Yet what sort of radiation could this possibly be? Further experiments revealed that these rays didn't appear to be some kind of invisible light rays. They didn't reflect off surfaces; and they didn't seem to exhibit refraction when they passed from one medium into another, the way light waves are bent when they pass from air into water. (He was in fact mistaken here.) Röntgen was baffled as to the nature of these mysterious rays, so he gave them the mathematical symbol for an unknown quantity – calling them x-rays.

Röntgen realized at once the sensational nature of his discovery. These x-rays could actually see *through* things – even Jules Verne hadn't imagined such a thing. Yet he knew this discovery would not remain uniquely his for long. So much research was being done into luminescence, someone else was bound to come across these x-rays sooner or later.

Keen to establish his priority, Röntgen embarked upon a speedy but exhaustive investigation into the properties of the newly discovered rays. He found that they could pass through paper, wood, and even thin sheets of metal. Other properties included their ability to ionize gases, and yet they were unaffected by electric or magnetic fields. They may have been invisible, but they affected photographic plates – which meant you could obtain photographic evidence of their passing.

After seven weeks Röntgen was ready to reveal his findings. In January 1896 he announced his discovery at a public lecture. The climax of this lecture came when Röntgen invited the venerable Swiss anatomist Rudolf von Kölliker to step up from the audience. Röntgen then took an x-ray photograph of the 80-year-old man's hand. When this was seen to reveal the entire bone structure of von Kölliker's fingers and wrist, the audience jumped to their feet and burst into spontaneous applause.

News of Röntgen's sensational discovery quickly spread all over Europe, and across the Atlantic to America. X-rays were so simple to produce that in no time they were being put to practical use. Just four days after Röntgen's discovery reached America, x-rays succeeded in showing the position of a bullet in a patient's leg. Dramatic stories began appearing in the press about the amazing properties of these new rays. The state of New Jersey even considered passing a law forbidding the use of x-rays in opera glasses, in order to safeguard the virtue of women attending the theatre. But alas, no one else thought of protecting the public from x-rays. Several years were to pass before it was discovered that over-exposure to x-rays caused leukaemia.

The almost accidental discovery of x-rays by Röntgen on 5th November 1895 is now regarded by some as the beginning of the Second Scientific Revolution. (The First Scientific Revolution began with Copernicus and his discovery that the Earth travels around the Sun, and came into its own with the scientific method of Galileo.) Though Röntgen himself was not aware of this at the time, his discovery meant that the age of classical physics (the mechanical world of Galileo and Newton) was coming to an end.

Initially, a move was made to call these new rays Röntgen rays. But confusion arose over the correct pronunciation and spelling of his name. (This confusion still exists in biographical dictionaries, where the Americans tend to list him as Roentgen.) And besides, the press found the name x-rays far more exciting. But Röntgen still achieved widespread recognition, and was awarded the first Nobel Prize for physics in 1901. However, some unkind French and British critics maintained that Röntgen was simply a run–of–the–mill experimenter who just happened to hit lucky. Others, mainly from Germany, claimed he was one of the greatest experimentalists of his time. Either way, he was undeniably possessed of one exceptional quality. Röntgen refused to patent anything to do with the production or use of x-rays, believing that these should be used for the benefit of humanity.

His discovery could have brought him immense financial reward. Yet he was to die old and impoverished in 1923, after his life's savings had vanished in the German hyperinflation (when a loaf of bread eventually cost millions of marks).

The next step in the new scientific revolution came as a direct result of Röntgen's discovery. This was made by the French chemist Henri Becquerel, who came from a well-established scientific family. His grandfather had fought at the battle of Waterloo, and later pioneered electrochemistry. His father had continued in this field, investigating fluorescence and phosphorescence. These are the phenomena which occur when matter absorbs light of one wavelength and emits it at another. (Perhaps the best example of this occurs when invisible ultraviolet light is shone on different minerals making them glow in different colours.)

When Becquerel heard of Röntgen's discovery of x-rays, he was reminded of his father's experiments in fluorescence. Röntgen had discovered x-rays by noticing the fluorescent effect they had on barium platinocyanide. This made Becquerel wonder whether any fluorescent materials themselves produced x-rays.

Early in 1896 Becquerel began experimenting with a double salt of uranium (potassium uranyl sulphate), which he knew from previous work was capable of high fluorescence. He put a crystal of this salt on a photographic plate which had been wrapped in black paper, and then placed these in the sunlight. He knew the sunlight would induce fluorescence, and if this contained x-rays they would penetrate the black paper and register on the photographic plate. This was precisely what happened. When Becquerel unwrapped the photographic plate, and developed it, he found a faint white fogginess spreading out from around where the crystal had been placed. There was only one conclusion. Fluorescence produced x-rays!

It was grey winter in Paris, and sunny days on which he could press ahead with his experiments were few and far between. Impatiently Becquerel occupied his time preparing further photographic plates, wrapping them with black paper and placing the crystal on top. These he carefully placed in a darkened drawer. But still the sun didn't shine.

Unable to contain himself any longer, Becquerel decided to have a look at a couple of the photographic plates in the drawer, just to see if perhaps the crystals might have emitted some remnant faint luminescence. He was in for a shock. When he developed the first plate he discovered an intense white fogginess spreading out from where the crystal had rested. This meant that whatever radiation was being emitted by the salt did not involve sunlight. And it didn't involve any visible type of fluorescence either, as the crystal hadn't been glowing in the dark.

Becquerel immediately began investigating this unexpected radiation. He found to his surprise that it wasn't quite the same as x-rays. Could this be an entirely new form of radiation? Like x-rays it was invisible, and could ionize gases (leaving an electric charge in the air through which it passed). But it could penetrate matter far more strongly than x-rays. And he also noticed a much more curious effect. The potassium uranyl sulphate crystal continued to give off a constant stream of this radiation. This didn't appear to depend upon whether it was placed in the light or the dark. It just radiated continuously in all directions.

At this point Becquerel's investigations came to a full stop. These phenomena were all very interesting, but they didn't seem to be leading anywhere. Becquerel was an experimentalist, not given to imaginative leaps of theory. Though ironically, his understanding of this new type of radiation was fatally hampered by theoretical preconceptions which he had inherited from his father. Although the radiation didn't appear to be caused by the Sun, or by light, he remained convinced that it must be a 'type of invisible fluorescence'. What else could it be? Radiation couldn't just stream out of crystals continuously without there being some source of input. Perhaps this input could somehow be stored by the crystal over a period of time?

Becquerel's was the correct approach. This was how the laws of classical physics worked. For over 2,000 years, since Ancient Greece, scientific thinkers had held to a literal interpretation of the saying attributed to Epicurus: *Nihil ex nihilo fit* ('Nothing is made out of nothing.') But science was changing. Becquerel had made a highly important discovery, but it was not a new form of fluorescence. So what was it?

Marie Curie had followed the discoveries made by Röntgen and Becquerel with great interest, as ever discussing them with Pierre. She had by now finished her research on magnetism, and was looking about for a suitable subject for her doctoral thesis. Becquerel's dead end in his experiments offered an exciting challenge. Marie decided to study this new form of radiation.

Marie Curie had now finished her stint as a member of Lippmann's research team, which meant she no longer had access to the new labs at the Sorbonne with all the latest equipment. Instead, Pierre obtained permission for her to take over some old storage space in his labs at the School of Industrial Physics and Chemistry. This was little more than a chilly, drab corner space, and she had to start equipping it from scratch. But here Marie Curie had something she would never have been able to obtain at the Sorbonne labs: complete autonomy. She could follow her research as she pleased, wherever it led.

According to her laboratory notebooks, Marie Curie began her experiments here on 16th December 1897. She started studying the radiation given off by potassium uranyl sulphate, a repeat of Becquerel's experiment. Referring to this radiation in her notebooks, she coined the term 'radioactivity'. Like Becquerel before her, Curie confirmed that radioactivity 'electrified' the air through which it passed. The air became ionized, thus capable of conducting electricity. As the radioactivity became more intense, the ionization increased. Even so, the quantities to be measured were minute – of the order of 50×10^{-12} amperes. This required an extremely delicate measuring instrument.

For this Marie Curie was able to make use of the piezoelectric effect discovered by Pierre Curie and his brother Jacques just over a decade previously. Using the fact that a crystal under pressure emits a minute charge of electricity, she used this to counterbalance the opposite minute charge in the air through which the radioactive rays were passing. Thus a reading of zero electrical charge was achieved. So the greater the pressure on the crystal required to counterbalance the electrical effect in the radioactive air, the greater the radioactivity.

Marie Curie now began to study various

different uranium compounds, ranging from pitchblende to certain uranium salts. The black-brown pitchy ore pitchblende, a mineral form of uranium oxide, was found to be highly radio-active, giving a reading of 83×10^{-12} amperes. On the other hand, some of the uranium salts registered a mere 0.3×10^{-12} amperes. But in the course of these experiments Marie Curie made one important discovery. It didn't seem to matter if the compound was heated, in solution, or in powdered form. Only one thing affected the amount of radioactivity – the amount of uranium present. The source of the radioactivity was not uranium compounds: this was a property of uranium atoms themselves.

But was this property unique to uranium? Marie Curie began some experiments with atoms of similar atomic weight. Thorium oxide produced ionization requiring a piezoelectric charge of 53×10^{-12} amperes to neutralize it. Uranium was not unique in this property: thorium too was radioactive.

But these weren't the only important dis-coveries Marie Curie had been making. 'Two uranium ores,' she explained in her report, 'are much more active than uranium itself [which] leads one to believe these ores may contain an element even more active than uranium.' For instance, uranium ore from pitchblende gave a reading four times higher than accounted for by the amount of uranium it contained. There seemed to be no way of accounting for this, unless pitchblende contained another radioactive element. But this would have to be present in minute quantities, otherwise its presence would already have been detected. And it would have to be *extremely* radioactive, to account for the high overall radioactive readings. Also, as no other element had been found which contained anything approaching such levels of radiation, it was likely that this would be a hitherto unknown element.

Marie Curie's bold scientific thinking looked as if it was leading her towards a major discovery. As always, Marie and Pierre had discussed the progress of their work together in their evenings at home. In this way, they had both contributed towards each other's work. But Pierre now realized that his wife's work was becoming of major importance. As a result, he decided to abandon his own research entirely, and join Marie in hers. According to the legend, Pierre Curie knew that he was a brilliant scientist, but now realized that his wife was on the point of becoming one of the all-time greats. It was obvious who was the major player in the Curie partnership. As we shall see, such one-sided accounts are very much open to question. In fact, there were two great scientists at work here. Already they had established a remarkable rapport in their professional and emotional lives (two realms which were far from separate), so Pierre's decision to abandon his own research, and throw in his lot with Marie, was not quite so drastic as it might at first appear.

Another point worth bearing in mind is that throughout this period of intense experimental work, Marie Curie was at the same time bringing up their young daughter Irène – born just three months before she made her first entry in the experimental notebook she used at Pierre's labs. Marie employed a maid to help her look after Irène, but it has been claimed that she never missed giving the baby her evening bath. This too may well be part of the legend. On the other hand, we now know that Marie and Irène Curie went on to become arguably the greatest parent-child partnership ever seen in science. The psychological foundation of this relationship must have been laid during the first five years of baby Irène's life – the period of Marie Curie's most intense research. All this makes Marie Curie's achievement even more remarkable. She may have been an intense and highly focused personality, but she also managed to achieve a remarkable balance. She was no genius lost in a world of her own. This was a mind functioning at the highest level in an environment of nappies and wailing in the early hours. (Curiously, just a few years later Einstein too did his greatest work in similar circumstances – though being a male of his era, he would not have been as intimately involved as Marie Curie with the actual poo and *waaa* . . .)

Together Marie and Pierre Curie now began the difficult task of trying to discover the unknown element in pitchblende. First they had to try and isolate the unknown element, which was only present in minute quantities. This involved refining the ore by chemical treatment and repeated distillation until they managed to obtain a sample of the element itself. But it proved impossible to isolate the element from the almost identical bismuth in the ore. By July 1898 they had obtained a few specks of bismuth powder which contained the new element. In the words of the Curies' joint report, this powder contained 'a metal not yet determined, similar to bismuth'. They added: 'We propose to call it Polonium, from the name of the homeland of one of us.'

Polonia is the latinized form of Poland. Individuals, planets and even a dog, all have elements named after them (Einsteinium, Uranium, Plutonium). Poland is one of the few countries to have achieved this distinction. It did so at a very necessary time, when the name Poland was in danger of disappearing from the map. Marie Curie may have emigrated to France and married a Frenchman, but she was to remain throughout her life a patriot, and intensely Polish. Her spoken French, for instance, though fluent, was always spoken with an unmistakable Polish accent.

The discovery of Polonium was announced by the Curies in a joint paper entitled 'On a new radio-active substance contained in pitchblende'. (This was the first time the word 'radio-active' appeared.) They found that polonium was 400 times more radioactive than uranium. Yet even this high level didn't account for the level of radiation found in pitchblende. There appeared to be another highly radioactive element present. Once again they began hunting for the needle in the haystack. This time they managed to isolate an unknown element in some specks of barium powder, and felt able to announce: 'We have found a second radioactive substance, entirely different from the first in its chemical properties.' This element could only be differentiated from barium because of its high radioactivity.

In order to clarify matters they called in the chemist Eugene Demarçay, an expert in the new field of spectroscopy. (This involves the use of a spectroscope, which converts the light emitted by a substance into a spectrum. Every substance has its own characteristic spectrum, whose lines indicate its chemical properties.) Despite losing an eye in a laboratory accident, Demarçay had become extremely skilled at reading the complex line patterns of spectra. Yet even he was at first unable to identify any new spectral lines in the tiny sample of barium which the Curies claimed contained their second new element. But they knew it was there, because of its high radio-activity. Its lines were obviously almost identical to those of barium. After repeated attempts Demarçay finally managed to detect, amongst the barium lines, a few similar but incontrovertibly new lines. The Curies had indeed discovered another new highly radioactive element – which they named 'radium'.

The Curies were determined to examine the properties of this remarkable new element, which appeared to emit a continous stream of intense energy without diminishing itself. But in order to examine radium, they were going to need a vast amount of pitchblende. Only by starting with almost industrial quantities of this ore would they be able to produce radium in sufficient quantities for them to determine its atomic weight and analyze it. But where could sufficient pitchblende be found? The Curies made enquiries and heard of a mine at St

Joachimsthal in Bohemia (then part of the Austro-Hungarian Empire, now in the Czech Republic). The mine here produced silver and uranium, but the waste ore from which the silver and uranium had been extracted contained pitchblende. The mine was literally surrounded by slagheaps of ore bearing hidden traces of radium. The mine owners were only too pleased to get rid of this worthless stuff to these two crazy French scientists. As long as the Curies were willing to pay the freight charges, they could have as much as they wanted.

(By coincidence, this mine was again to play a significant role in the history of science some 40 years later. When Hitler took over Czechoslovakia in 1938, a list was issued of substances banned from export. Tucked away in an inconspicuous part of this list was uranium from the Bohemian mines. When the Danish scientist Niels Bohr noticed this, he realized that the Nazis had begun serious research into making an atomic bomb. He warned the Americans, who immediately embarked on the process which led to the first nuclear bomb.)

In order to transport the pitchblende from the

St Joachimsthal mines to Paris, the Curies were forced to dig deep into their meagre savings. But where could they find the space to process such a vast quantity of ore? This time Pierre Curie secured permission for himself and his wife to take over a large disused shed in the grounds of the School of Industrial Physics and Chemistry.

The shed had formerly been a dissecting room, but now its grimy glass roof leaked onto the cracked concrete floor. The place was freezing in winter and stifling in summer. In the words of a colleague, it was a 'cross between a stable and a potato cellar'.

Here Marie Curie began the colossal task of reducing mounds of pitchblende waste into specks of radium powder. Each stage in this process was meticulously recorded in her laboratory notebook. Yet these columns of figures are not just the dry-as-dust record of a long and laborious process. Achievements are liable to be greeted with an excited barrage of exclamation marks. 'March 14th. Precipitate in cylinder 4.3!!!!!!!!!!' Such entries bring her notebooks to life – giving a glimpse of the human element. This was a brilliant woman passionately engaged in a labour of love. Such moments were bliss to her.

But this wasn't Marie Curie's only labour of love, or even her only notebook. At home she had been keeping another notebook, filled with scientific observations of a different nature. These recorded the progress of her daughter Irène. With due solemnity the child's weight and 'length' were regularly recorded. Even the diameter of her head was measured with pincers. Each new stage in the experiment was written up in exemplary scientific fashion. Thus we learn that in July 1898 Irène said: 'gogli, gogli, go.' 'On August 15th . . . Irène has cut her seventh tooth.' And: 'On January 5th 1899: Irène has fifteen teeth!'

Indeed, Madame Curie seems to have become something of a notebook freak. Her home notebooks record a wide variety of Curie family activities – including a recipe for gooseberry jam, a draft of a letter to the Academy of Science (informing this body of the discovery of radium), a report of Irène singing, and long meticulous lists of household accounts. From these we learn the cost of the material for Pierre's shirt, the freight charge for pitchblende from St Joachimsthal, the wages for the maid-cum-child-minder.

Pierre's salary as head of the laboratory at the School of Industrial Physics and Chemistry was not high, and the Curies had to struggle to make ends meet for at least the first five years of their marriage. Even so, their circumstances weren't quite so straitened as the legend would have us believe. The Curies worked long hard hours in their primitive lab-shed, and didn't indulge themselves at home. But they still managed to cycle out into the countryside on sunny weekends (while Pierre's family looked after Irène.) And like all others who could, they usually fled the hot smelly streets of Paris during the summer flea season – spending a long holiday in the country, where the living was as cheap as it was idyllic. In 1898 there is a three month gap in Marie Curie's laboratory notebook. This was during the *grande vacance*, when the Curies were on holiday in the Auvergne. Here the local wine flows like water, the goat's cheese is out of this world, and rivers wind through remote mountain valleys, with secluded rock pools where it is safe enough to bathe with a child whose skull has a diameter of precisely 12.4cm. (Prior to going on this particular holiday the Curies had just discovered polonium. When they came back they discovered radium. For many, such 'straitened circumstances' are the stuff of dreams.)

Despite the occasional outburst of exclamation marks in Marie Curie's laboratory notebook, the task of extracting molehills of radium from mountains of pitchblende was no easy matter. The method she eventually chose to use was long, tedious and exacting. She was in fact embarking upon an industrial process from scratch, and virtually single-handed. (Indeed the procedure which Marie Curie invented was the one later to be used by industry.) The pitchblende waste arrived from the mine in bags of brown dust mixed with pine needles. (The slagheaps around the mine were in pinewoods.) This was then dissolved in a chlorine solution, from which radium-bearing barium precipitated in chloride form, and could be filtered out. Radium chloride is marginally less soluble than barium chloride. So the mixed chlorides were then subjected to successive crystallizations, each crystallization producing a slightly richer concentration of radium.

'I had to work with as much as 20 kilograms of material at a time,' she wrote, 'so that the hangar was filled with great vessels full of precipitates and of liquids. It was exhausting work to move the containers about, to transfer the liquids, and to stir for hours at a time, with an iron bar, the boiling material in the cast iron basin.' The fractional crystallization process, on the other hand, was a subtle task. 'The very delicate operations of the last crystallizations were exceedingly difficult to carry out in the laboratory, where it was impossible to find protection from the iron and coal dust.'

Yet these long days working in the 'hangar' beside Pierre were a time of selfless joy. 'We were very happy in spite of the difficult conditions under which we worked.' They would eat a 'simple student's lunch' amongst the apparatus. 'A great tranquillity reigned in our poor shabby hangar; occasionally, while observing an operation, we would walk up and down talking of our work, present and future. When we were cold, a hot cup of tea, drunk beside the stove, cheered us. We lived in a preoccupation as complete as that of a dream.'

The Curies' partnership was so close that it's often impossible to separate their different roles. Perhaps the most accurate reflection of who exactly did what is seen in the laboratory notebooks. These indicate that during the actual discovery of radium (and polonium) their work was virtually interchangeable. Lines of Marie's neat writing are interspersed with the inky spider-walk of Pierre's script. After discovering the two new elements they continued to work side by side, but on separate tasks. Marie took on the chemist's role of extracting radium, while Pierre used physics to investigate the nature of its radioactivity. But as we shall see, even at this stage their roles were not entirely separated.

Any attempt at crude psychological generalizations about the Curies' partnership is out of place. At this point Pierre did admittedly take on the more 'male' abstract task. And Marie did busy herself with the more practical task of cooking up radium in her hell's kitchen. But such sexist stereotyping is inevitably superficial where the Curies are concerned.

Prior to this separation into Mr Radioactivity and Mrs Radium, things had looked very different. Marie had shown that she was Pierre's superior at mathematics, and Pierre's invention of the delicate quartz instrument for measuring piezoelectricity had proved his brilliant practical skills. Admittedly, after Marie embarked upon her single-minded endeavour to isolate radium, Pierre is unlikely to have got his hands dirty stirring the cauldrons and bubbling cast iron basins. But although he had little or no input into Marie's activities, the opposite was certainly not the case.

After baby Irène had been given her daily bath, weighed and measured, and put to bed, the Curies still had their evenings together. And as always, their absorbing topic of conversation was their work. (Not surprisingly, visitors to the Curies' sparsely furnished house were rare.) There wasn't all that much you could say about Marie's work on a daily basis. ('Crystalline deposit 3.2!!!!!!!!!') On the other hand, Pierre's work was just the kind of thing they were used to discussing together, as equals, with each making vital contributions.

In the course of this work Pierre (and Marie) Curie made important advances at the very threshold of scientific understanding. Pierre set up an experiment in which radioactive radiation passed through a magnetic field. He found that it separated into three different types of rays – alpha rays, beta rays and gamma rays (as they came to be called).

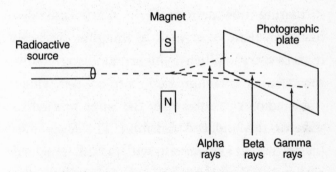

At the same time as Pierre Curie was making this discovery, it was also being made separately by Becquerel, and by the New Zealand-born physicist Ernest Rutherford (who was responsible for naming the rays). It was Pierre Curie who discovered that beta rays were negatively charged – meanwhile Rutherford established that alpha rays were positive and gamma rays were neutral.

As we can see, the work that the Curies were doing separately, yet together, had echoes in the work that Rutherford and others were doing separately, yet together. All involved in this work sensed that it was of the utmost importance, yet no one quite understood what these remarkable discoveries meant. That would only come with hindsight.

During this period Pierre Curie also discovered there was such a thing as 'induced radioactivity'. When a highly radioactive substance such as radium came into contact with a non-radioactive substance, the latter seemed to take on this 'induced' radiation. The equipment he used in his experiments with radium remained radioactive long after the experiment was over and the radium removed.

Pierre began comparing notes with Becquerel, whom he had known for some years. Becquerel had now overcome his impasse after the discovery of radioactivity, and was making considerable experimental progress himself. He had found that when he carried even the tiniest quantity of radioactive material in his pocket, it produced a burn on his skin. Pierre used a miniscule sample of Marie's radium and came up with the same result. He then undertook further experiments, and discovered that one gram of radium gave off 140 calories per hour – sufficient to boil water! Not only was radium a very strong source of energy, but the implications of this energy were sensational. In the words of the Curies' report: 'Every atom of a radioactive body functions as a constant source of energy. . .which implies a revision of the principles of conservation.' (When this news was to become public a few years later, the press ran the headlines: 'Curies Discover Perpetual Motion!' And for all science could say, they were right.)

No one knew what was happening. No one knew the cause of all this. But similarly, no one knew the effect of all this either. And this was to have disastrous consequences. Pierre Curie's experiments on his skin with radium had produced radiation burns. The 'induced radioactivity' he had discovered in his experimental equipment was the very radioactive contamination which we now so fear. Much as with x-rays, it was to be some years before anyone realized the dangers of radioactivity. During the four long years that Pierre and Marie worked together on radium in their 'hangar' on the Rue Lhomond, neither of them thought to take the slightest protection against radioactivity. The extent of this can be judged from the fact that Marie Curie's famous laboratory notebooks were so contaminated with radioactivity that they remain too dangerous to handle *to this day* ... However, there was apparently little effect on Marie herself at the time – though this was not long to remain the case.

Radioactivity was leading science into a new era. Classical physics, as epitomized by Newton, believed that the universe worked in an essentially mechanical fashion. Vital to this world-view was the idea of matter. The ancient Greek philosopher Democritus had put forward the idea that ultimately matter consisted of small indestructible 'atoms' (which is Greek for 'indestructible'). But this materialist view proved far too progressive for the fifth century BC, and was soon being dismissed by such leading philosophers as Plato and Aristotle – thus consigning it to oblivion for over two millenia. But in the end the truth will out. By the 17th century an idea identical to Democritus's atom was making a long-overdue comeback. Though when it was first introduced into classical physics the word 'atom' was not used. Newton chose to describe this atom-in-all-but-name as 'solid, massy, impenetrable movable Particles . . . even so very hard, as never to wear out or break into pieces'. Such atoms were seen as acting like minute billiard balls.

By the end of the 19th century, when the Curies were at work on radioactivity, this view had become well established. However, there was a plausible school of scientific (and philosophical) thought which had begun to question the very existence of atoms. According to them, science progressed by hard-nosed practical thinking, which was backed by experimental evidence. In line with such rigorous empirical thinking was the Viennese theoretical physicist Ernst Mach, who believed that ultimately all knowledge is derived from sense data. Science was refined experience, and nothing else. In his view, science had inherited a certain amount of historical baggage which didn't conform to these standards. Amongst this was the idea of the atom. Who had ever seen an atom? Science didn't need to believe in little invisible (and indeed indestructible) billiard balls to explain what was happening in reality. It was just an idea which science had inherited from previous unscientific thinking.

Most scientists of the time took Mach's views with a pinch of salt. Science was more than just experimental data; it would always need ideas of some sort. The fact was, atoms conformed to that other great scientific notion: if it works, use it. Ideas too can produce results – which can of course be tested by experiment. As an explanation of the ultimate nature of matter, atoms appeared to work. They aided further thinking in both physics and chemistry. Using this notion had helped to discover things, been fruitful in producing useful theories, and so forth. (In 1871 the Siberian-born chemist Dmitri Mendeleyev virtually reinvented chemistry when he published his Periodic Table of the Elements. This listed all chemical elements according to their atomic weight and valency (ability to combine with other elements). The fact that he had never actually seen an atom doesn't seem to have bothered him overmuch.)

But it was Mach's questioning of atoms which introduced the first crack into the billiard ball concept of the atom. At the same time as the Curies were investigating radioactivity in Paris, the physicist Rutherford and the eccentric chemist Soddy were doing the same thing on the other side of the Atlantic in Canada. The bluff New Zealander Rutherford was a good foil for Soddy, whose hobby was preaching a new world economy where money was banished. This brief partnership was to produce a major breakthrough. The approach of Rutherford and Soddy was more theoretical than the Curies'. In 1902 Rutherford and Soddy published a paper entitled 'The Cause and Nature of Radioactivity'. This stressed an essential difference between x-rays and radioactivity, which had hitherto appeared remarkably similar. X-rays were produced when a substance was bombarded, whereas radioactivity occurred spontaneously. Rutherford and Soddy decided that radioactivity was definitely an atomic phenomenon. It appeared to be a form of atomic decay, whereby certain unstable heavy atoms disintegrated to become more stable lighter atoms. Their famous conclusion about radioactivity was that 'these changes must be occurring within the atom'. The solid billiard ball fell apart. The era of classical physics was giving way to the new age of nuclear physics (which is concerned with the 'nucleus' of the atom ie, sub-atomic particles). Rutherford and Soddy had put forward the radical notion that atoms could change. The Curies had shown that radioactivity was a colossal, and apparently constant energy source. (They were in fact slightly wrong about the latter.) But still no one really knew what was going on.

The first big step in this direction was to come a couple of years later. An obscure Swiss amateur physicist, who worked in the Patents Office at Bern, came up with a literally earth-shattering idea. In 1905 Einstein published his paper on the special theory of relativity. This led him to derive the most famous formula since Pythagoras's theorem:

$$e = mc^2$$

In other words, a tiny amount of matter (m) could turn into a colossal amount of energy (e); the constant (c) was nothing less than the speed of light (around 300 million metres per second)! We now know that this is what produces radioactivity. What Becquerel and the Curies had discovered would one day lead to nuclear energy (and nuclear bombs).

By early 1902 Marie Curie had at last succeeded in producing some radium. This had required processing over a tonne (1,000 kilograms) of pitchblende waste at the rate of 20 kilograms a time. The crystallization and re-crystallization process had needed to be repeated many thousands of times. And in the end she had managed to produce one tenth of a gram of radium. (As this process ground on, the barrages of exclamation marks in Marie Curie's laboratory notebooks gradually diminished, only to reappear with a vengeance towards the end.) This miniscule quantity had been enough for Demarçay to identify the spectrum of radium and establish its atomic weight – thus dispelling once and for all any lingering doubts about whether radium really was an element. Eventually Marie

Curie was to process almost ten tonnes of pitchblende waste in her hangar – from which she finally managed to extract a whole gram of radium.

Despite all this immense effort, and the great commercial possibilities for radium, Marie Curie refused to patent her method for producing radium from pitchblende. Heedless of their poverty, the Curies had decided together that the benefits of radium should be made available to the world. Such a patent could have brought her untold riches. Marie Curie had exceptional qualities, which were not limited to her scientific abilities. The pity is, her later biographers insisted on inventing a few more qualities (and over-looking a few frailties, too) – until the legend quite obscured the human being. As we shall see, Marie Curie was no legendary figure: she always remained essentially human.

By 1903 Marie Curie had written up her heroic attempt to run an entire industrial production process single-handed. (There were lab assistants, and the likes of Becquerel and Demarçay were consulted in matters relating to their respective fields. But the bulk of the work was done by a 33-year-old mother who insisted on bathing her daughter at night, and enjoyed nothing more than sorting out a few of her husband's theoretical problems after dinner.)

Marie Curie's doctoral thesis earned her the first advanced research degree awarded to a woman in France in any subject. Its exceptional merits were speedily recognized throughout the world. Later in the same year Marie and Pierre Curie, together with Henri Becquerel, were awarded the Nobel Prize for Physics.

Ironically, Marie Curie's input had of course been in the field of chemistry. Misclassifications of this order are not so rare as one might expect from such an august intellectual body as the Nobel committee, which is backed by a worldwide network of expert advisers. Perhaps the most notorious of these mis-awards was to Henry Kissinger, the US secretary of state during the Vietnam War, who was given the Nobel Peace Prize. (Though the committee did draw the line at Winston Churchill's request for a Peace Prize, deciding to give him the award for literature instead.)

Neither Pierre nor Marie were able to attend

the presentation of the Nobel awards by the King of Sweden in Stockholm. Pierre had recently managed to pick up some lecturing work, a vital addition to the family income, and decided he just didn't have time to go and pick up a Nobel prize. Marie, on the other hand, was a little ill, and had no intention of leaving her daughter in the hands of someone who didn't know how to bath her properly, let alone measure the diameter of her skull. Alas, Marie's illness on this occasion was almost certainly the first signs of radiation sickness.

The Nobel award to the Curies caught the public imagination. ('Man and Wife Team Discover Perpetual Motion in Hut' etc.) This meant the Curies now had to put up with all the disruptions of popular fame, a phenomenon which was already entering the full flush of fatuity by the beginning of the 20th century. The sheer intrusion and banality of it all came as quite a shock to a serious-minded woman such as Marie Curie.

So there was little time for serious research in 1904. But this wasn't all because of newspaper interest, postbags of fan mail, total strangers banging on the lab door, and the like. As a direct consequence of the Nobel prize Pierre was appointed to the newly-created professorship of physics at the Sorbonne. By now Marie had managed to obtain a part-time teaching post at the female branch of the Ecole Normale Supérieur at Sévres, on the outskirts of Paris. She was the first woman on the faculty there, despite the fact that it was an all-female college (the most prestigious in French higher education). The lack of any full-time appointment for Marie Curie was only partly due to discrimination. In the summer of 1904 Marie had a miscarriage, a traumatic event which laid her low for some months. It wasn't until Marie was 38, a venerable (and dangerous) age for childbearing in those days, that she successfully gave birth to her second daughter, Eve, in December 1905.

By now radium was in large-scale production. Gambling on the commercial profits to be made from this newly-discovered wonder, an enterprising French manufacturer Armet de L'Isle had erected a factory for its production in the suburbs of Paris. Marie had freely given him full details of her radium-manufacturing process, and refused to accept any payment for her invaluable assistance. De L'Isle was pleased to accept this, but was not so pleased when competing foreign manufacturers also availed themselves of this free advice service.

By 1906 both Marie and Pierre Curie were beginning to show signs of what is now recognized as radiation sickness. This disease was unknown at the time (for obvious reasons), so no one really knew how to treat it. Yet in the light of what we now know, the astonishing fact is that they were still alive at all. At this stage Marie and Pierre were both suffering from severe radiation burns on their hands. Pierre was also afflicted by severe pains in his arms and legs, to the extent that he was beginning to find it difficult to dress and undress himself. Then in April 1906 disaster struck.

After a hard day's work at the Sorbonne, Pierre set off home along the narrow Rue Dauphine in the Latin Quarter. It was pouring with rain, and he was forced to huddle under his umbrella. At one point he absent-mindedly stepped off the pavement – right into the path of a six-tonne horse-drawn wagon. He was knocked down and fell under the wheels. In the words of his daughter Eve Curie, vividly describing the event many years later: 'The rear left wheel of the cart encountered a feeble obstacle which it crushed in passing: a forehead, a human head. The cranium was shattered and a red, viscous matter trickled in all directions in the mud: the brain of Pierre Curie!'

Marie Curie was devastated. Her husband, her co-worker, the only mind of her own calibre with whom she had been able to establish an immediate and close rapport – all gone! For some time afterwards, when making entries in her laboratory notebook she continued in her habitual fashion: addressing him by name and asking him questions. Her first paper after his death even begins: 'Pierre Curie observed some years ago . . .' But she didn't make a cult out of him: Marie Curie had no religion, and was not intent on making one out of her intense grief.

A few months after Pierre's death, Marie was appointed in his place as professor of physics at the Sorbonne. This was an unprecedented step for a woman – even a world-famous Nobel prizewinner. (The only logical inference here is that being a woman was regarded as a *physical* defect, which somehow left half the human race *invalid*: no matter what you did, it just didn't count.)

On 5th November 1906 Marie Curie stepped up to the lectern to deliver the first lecture given by a woman at the Sorbonne in its 600-year history. She began with no preamble – launching straight into the last words which Pierre had delivered in his last lecture, and continuing seamlessly with the same subject. Her voice was weak, and her delivery montonous, but her intensity of manner soon gripped the audience. This was advanced science being taught by someone who was actually at the forefront of that advance: the latest news from the front. (Coincidentally, it was 11 years to the day since Röntgen had made the fortuitous discovery of x-rays which had set the ball rolling on 5th November 1895.)

According to the hagiography written years later by her daughter Eve, Marie Curie now assumed a role on a higher sphere: she became something of a secular saint of science. Her prematurely aged face was haggard, her hair greying. This bleak demeanour was accentuated by the plain black dresses she wore. Such was the traditional widow in mourning. But the distinguished woman professor in Queen Victoria role was also a dedicated mother, dutifully bringing up her two young children. Eve was just four months old at the time of her father's death, which she describes in such colourful detail. Irène was by then nine.

But this is only half the story. Marie Curie was a passionately intense woman, not a Victorian effigy. Another picture from these years shows a saddened, rather than a haggard woman. Her upswept glossy hair has no flecks of grey, and she is wearing a fashionable, but restrained frilly white blouse. A personable, if not exactly merry, widow.

Marie Curie's laboratory work now brought her into contact with Paul Langevin, a former student of Pierre's. Langevin had a scientific mind of the highest order, and was well on the way to becoming the leading physicist in the land. He had done work in Cambridge with Rutherford's colleague J J Thomson, and later undertaken research into weak paramagnetism. He rightly surmised that this was due to the electric charge of subatomic particles. As we have seen, Marie Curie too had worked in magnetism; and had herself played a leading role in the birth of nuclear physics.

But Langevin was more than just a pretty brain, he also had a superb military moustache, complete with long side-wings waxed to curly upturned points. (Such magnificent accoutrements to male pride have sadly long been ridiculed into extinction.) Unfortunately Langevin also had a very fierce wife, and an even more fierce live-in mother-in-law. There were disagreements – resulting in frequent domestic violence. But in those days this was considered quite acceptable. Despite having 'natural male superiority' (to say nothing of his valiant moustache), the exceptional scientific mind proved no match for his two determined adversaries. They were wont to arm themselves, in most ungentlemanly fashion, with parasols, bottles or even a wooden mallet (intended for tenderizing the husband's steak rather than his skull). Yet despite living in this atmosphere of domestic high farce, and regular threats of divorce all round, by 1909 Langevin's marriage had miraculously produced four children in five years.

Not surprisingly, Langevin presented a somewhat sombre (and sometimes rather scratched) figure in the Sorbonne physics labs. But the combination of magnetism and moustache eventually proved too much for the new Professor of Physics. Langevin and Curie seem to have established a rapport which went deeper than scientific understanding some time in 1908. There are things that the eyes betray and instincts alone recognize, long before their possessors are willing to accept such evidence. It was at least a year – of constant professional contact, scientific dialogue of the highest order, and deep intellectual perception – before Marie and Paul could admit to each other what was happening. They were deeply and madly in love with each other.

In July Marie and Paul rented a small apartment near the Sorbonne. Colleagues would notice them exchanging notes at work. But instead of technical information on the nature of paramagnetism ('Suggest electrons align in applied magnetic field') these two geniuses were passing information all too comprehensible to lesser mortals ('I am so impatient to see you').

Having been preceded by high farce, the affair continued to follow the rules of this quintessentially French genre. Langevin's wife became jealous. Frantic letters were exchanged, and intercepted by the wrong parties. Langevin's wife threatened to kill Marie Curie. Langevin, who was probably the best judge of such matters, took this threat seriously. (Though for some reason he doesn't seem to have feared for his own life.)

Yet as with all such dramas, in reality the suffering of those involved was obviously profound. Marie Curie may have been passionately intense by nature, but she was no emotional sophisticate. Judging from the little evidence we have, her private turmoil was considerable. She was deeply in love, fighting to win her man away from his family, while Paul dithered and agonized between his family and the woman he loved. How the other combatants (and clingers on, both large and small) felt we should only conjecture with compassion.

Such private agonies were hard enough to bear – but worse was to come. The affair went public when Marie Curie's intercepted letters were 'stolen' and found their way into the hands of the popular press. Marie Curie now saw her tenderest feelings exposed for the nation to enjoy over its breakfast coffee and croissants: 'My dear Paul, I spent yesterday evening and night thinking of you, of the hours we have spent together ... delicious memory ... good and tender eyes ... all the sweetness of your presence.' The letters were accompanied by the traditional orgy of editorial hypocrisy and prurient speculation. Doubt was even cast on Pierre's death. Could he have been pushed?

The worst attacks came from Gustave Téry, the right-wing anti-semitic editor of *L'Oeuvre*, a former schoolfriend of Langevin who bore a grudge. Téry railed against the goings-on at the 'German-Jewish Sorbonne'. The fact that Marie Curie was Polish, and not of Jewish descent, was irrelevant it seems. To appoint a female foreigner as professor of physics was an insult to French manhood. (The French word *chauviniste* means 'patriot'.)

Things became so bad that eventually Langevin challenged Téry to a duel. In a final act of high farce Langevin and Téry set off to confront each other, accompanied by their seconds and a doctor. Pistols were drawn at 11 am on a foggy November morning in the Bois de Vincennes, the park east of Paris. (In the old days such confrontations traditionally took place at dawn: this more civilized hour was presumably to accommodate the attending press.) The two combatants slowly raised their pistols at one another. According to Téry, at this moment he felt overcome by 'scruples about depriving French science of a precious brain'. Sadly, Langevin seems to have suffered from similar scruples about depriving French journalism of the opposite. Both Langevin and Téry refused to fire their bullets. The farce had reached its climax.

After this Langevin decided to return to his family, and the scandal gradually died down. But the damage was done. In an echo of her first love for Kazimierz, Marie Curie had been humiliated. She had fallen for a man who in the end had chosen his family above his love for her. But this time she had not only lost her man, she had also lost her reputation. As a consequence of the scandal in the press, Marie Curie's name had become blackened thoughout Europe.

Just before the scandal had broken in the press, Marie Curie had been awarded a second Nobel Prize. This time it was for chemistry, to honour her discovery (with Pierre) of the new elements polonium and radium. After her love letters appeared in the press, the Nobel committee wrote Marie Curie a letter explaining that she wouldn't have been awarded the prize if they had known about all this. The suggestion was obvious, and was echoed in the press to whom this letter was leaked. She was expected to surrender the prize, gracefully ('like a gentleman', as one paper put it). It was hoped she wouldn't embarrass the King of Sweden by turning up in Stockholm to collect her medal.

But Marie Curie was no gentleman, and had no intention of trying to ape one. Instead she wrote back to the Nobel committee, pointing out that 'the prize has been awarded for the discovery of radium and polonium. I believe there is no connection between my scientific work and the facts of my private life.' She'd called their bluff: the prize was not withdrawn.

Indeed if Nobel prizes were to be withdrawn on the grounds of extra-marital sex, there would now be huge gaps in the lists of winners – from notorious womanizers like Einstein and Schrödinger, to James Watson (of DNA fame) with his youthful penchant for Scandinavian au pair girls. But there remained one scapegoat. Langevin was never to be awarded the Nobel prize, despite being responsible for the modern theory of magnetism and the invention of 'sonar' (early radar). Also, it's worth recording that this great mind wasn't all moustache when it came to facing up to the difficulties of reality. After the German invasion of France in 1940, the 68-year-old Langevin was one of the few who continued publicly to oppose fascism, eventually being forced to flee for his life to Switzerland (his daughter was sent to Auschwitz, his son-in-law shot).

Marie Curie's award of a second Nobel prize for science in 1911 was not to be equalled until 61 years later in 1972 by the US physicist John Bardeen. This time Marie Curie decided she

would go to Stockholm to collect her medal.
When she arrrived back in Paris in December
1911, she collapsed and was rushed to hospital.
The strain of the scandal, her vilification by the
popular press, Paul's return to his family – it had
all been too much for her. But this turned out to
be more than just a nervous breakdown.
Throughout 1912 and 1913 Marie Curie suffered
a succession of debilitating complaints. Radiation
sickness was beginning to take hold. She would
never again recover the former robust health
which had sustained her through her long hours
working beside Pierre in their shed on the Rue
Lhomond.

1914 saw the outbreak of the First World War.
The western front was soon bogged down in 400
miles of trenches, stretching along eastern France
from the Swiss Alps to the North Sea. There
were massive French casualties. Marie Curie
dropped her radium research and began
experiments which eventually resulted in a
portable x-ray machine. She campaigned for
funds to equip an ambulance, and soon took her
portable x-ray unit to the front.

By 1916 Marie Curie was running a fleet of

ambulances, and had taken her driving test so that she no longer needed to rely on a chauffeur. This was a radical step. In those days men drivers outnumbered women drivers by over 500-1. (Male drivers may be interested to learn that as this imbalance was reduced over the years, it was accompanied by a curiously parallel reduction in the accident rate.)

In 1916 Marie Curie was joined by her daughter Irène, who was now 18 years old. Irène assisted her mother in teaching radiology classes to military medical personnel, enabling them to use the new x-ray units unsupervised at the front. This was the start of a mother-daughter collaboration that was to last the rest of Marie Curie's life.

Immediately after the war Marie Curie opened the Radium Institute in Paris. This was devoted to research into the uses of radium, and quickly became a world-renowned centre for nuclear physics and chemistry. Marie Curie became director of the Institute, and Irène acted as her assistant. Both took an active part in the research, though Marie Curie soon found she had less time for such projects.

Marie Curie had by now become a world-famous scientific figure, a sort of female Albert Einstein. In those days the Germans were still proud to lay claim to the Swiss-Jewish Einstein; not to be outdone, the French championed as their own the Polish Marie Curie. Nationalism had only recently entered the field of science, and was still in its infantile stage. But it was to develop fast. Within 15 years Hitler was dismissing 'Jewish Science': a successful shot in the foot for Nazi scientific research.

Just as Einstein used his fame to good effect championing liberal causes, Marie Curie became an emblem of independent womanhood. With her two Nobel prizes and the two daughters she had brought up as a single parent, Marie Curie became an inspiration to the generation of women born between the two world wars. No subject was closed. Women were capable of doing as well as (or better than) men in science. And to do this didn't mean you had to forswear a family. (Can it be an accident that Margaret Thatcher, Golda Meir and Indira Gandhi all initially chose to study science?)

In 1921 Marie Curie was invited to the

United States. When asked what she would like as a present from the president, she requested one gram of precious radium. This cost a massive $100,000, but the sum was soon raised by the women of America. In Washington, President Harding duly presented Marie Curie, in the company of her two daughters, with a green leather case containing a replica of the radium. The dangers of radioactivity were now being understood. But as a result of Marie Curie's researches at the Radium Institute in Paris, so were its beneficial effects. Radium was now being used in radium therapy (or curietherapy, as it was known). This involved various forms of exposure to miniscule amounts of radium – either through 'inhaling' its radiation, drinking 'irradiated liquid', being bathed in 'radium solution', or in some cases injection. Radium therapy was being explored as a treatment for a wide variety of illnesses, most notably cancer, arthritis and certain mental illnesses.

All of this treatment was in the early stages of development, and was not helped by press exaggeration of the 'Cure For Cancer

Discovered!' variety. During the 1920s radium so grew in the popular imagination that it came to be seen as the wonder cure for all ills. Marie Curie's name was inescapably associated with radium, and all this sensationalism only brought her further publicity. She found much of this tiresome, but was not entirely averse to the limelight.

Madame Curie, as she was now universally known, developed a unmistakably proprietary attitude towards radium. This was her discovery, her element. Discussions arose about establishing a radium standard. This was necessary to ensure equivalence in medical treatments, but most urgently for numerical agreements in the pursuance of international research. Marie Curie was all for the radium standard. Indeed, she had largely instigated discussion of this topic. But she insisted that it should be established by her, in her way. And that the actual standard radium itself should be kept by her at the Radium Institute in Paris. The international authorities called in Rutherford to mediate with the 'proprietress of radium'. Rutherford was aware that Curie could be a difficult woman, but he was also well aware of the chauvinist prejudice which she continued to encounter amongst the scientific community. Typical of this was the attitude of the leading US chemist Bertram Boltwood of Yale, who constantly referred to her as a 'plain darn fool'. Fortunately Rutherford had more respect for the double Nobel prizewinner. The two of them established a rapport, Marie Curie duly established the radium standard, and was then persuaded to hand over the sample to the International Radium Standard Committee.

In 1932 Marie Curie travelled home to Poland to open a new Radium Institute in Warsaw. Her sister Bronia was appointed director. Poland had been free since 1918, and the Sklodowskas were now a national pride. But Marie Curie returned to her own precious Radium Institute in Paris. Here she was intent upon establishing the largest radium stockpile in the world. This was required for research as well as the treatment of illness. Indeed, the rarity of radium was beginning to cause some tension between these two causes. Only when particle accelerators appeared in the 1930s, and were used to produce radium in larger quantities, was the conflict partially resolved.

By 1932 Marie Curie was 65. Despite her age, and the increasing onset of radiation-related illnesses, she still kept herself remarkably fit. On holidays she would go for long walks in the Alps, and enjoyed swimming. Back in Paris she was still capable of working in the laboratory beside her daughter Irène long into the night.

Irène was now a leading figure at the Radium Institute, and establishing herself as a scientist of international renown. In 1926 she had married Frédéric Joliot, one of her mother's assistants who was rapidly proving himself an exceptional experimenter. Irène Joliot-Curie, as she was now known, and Frédéric Joliot were very much in love, and like her mother before her Irène established an extremely close working partnership with her husband. He was the brilliant extrovert, while she had the shy intensity of her mother. Photos of them show a stylish happy young couple – Irène hadn't inherited her mother's dress sense.

In January 1934 Joliot and Joliot-Curie made the important discovery of 'artificial radioactivity'. This was a direct follow-on from the 'induced radioactivity' discovered by Pierre Curie. The Joliot-Curies found that when aluminium was subject to one type of radioactivity (alpha rays), it was liable to retain an alpha particle and emit a neutron. This made it unstable, and radioactive. Having started as aluminium, it was now transformed into an unstable isotope of phosphorous – which would eventually decay, as a result of radioactive emissions, into the stable element silicon. This was something very close to the dream of the ancient alchemists. It may not have been possible to transform base metals into gold, but it was certainly possible to turn *some* elements into others. On top of this, the Joliot-Curies drew the momentous conclusion: 'scientists, building up or shattering elements at will, will be able to bring about transmutations of an explosive type.'

Marie Curie received the news of her daughter's momentous discovery with quiet pride. This was precisely what she had trained her daughter for. Five months later, in June 1934, Marie Curie was taken from Paris to a sanatorium in the French Alps. Here she died on 4th July 1934 at the age of 66. The cause of death was leukaemia, which had resulted from overexposure to radioactive radiation. She had at last paid the full price for those long years producing radium in the hangar.

A year later the Joliot-Curies were awarded the Nobel Prize for their discovery of artificial radioactivity. But this story has a coda worthy of a Victorian novel. Years later, Irène Joliot-Curie's daughter married the grandson of Paul Langevin.

In 1938 Marie Curie's second daughter Eve published her official biography of Marie Curie, which became an immediate and inspirational bestseller, being translated into a score of languages. In this worshipful tome 'Madame Curie' was carved in monumental stone: a heroic figure who had given her life for science. Eve Curie's book relates many endearing and revealing personal details, which would otherwise have passed into oblivion, and for this we must be grateful. But this hagiography has certain glaring faults. Marie Curie was not a saint. The affair with Paul Langevin is passed over in silence; the agonies Marie Curie endured over the consequent public scandal remain unmentioned. This is an insult to the womanhood of the 20th century's most exceptional woman.

IMPORTANT DATES FOR MARIE CURIE & RADIOACTIVITY

1867	Born Maria Sklodowska in Warsaw
1878	Mother dies
1883	Wins gold medal on graduation from Russian Lycee in Warsaw
1883	Goes to live with uncles in the country for a year
1886	Becomes governess to Zorawski family, supports sister Bronia in Paris
1891	Goes to study in Paris
1893	Comes first in *licence* of physical sciences at Sorbonne
1893	Enters Lippmann's research laboratories
1894	Meets Pierre Curie, whom she marries the following year

1895	Röntgen discovers X-rays
1896	Becquerel observes radiation
1897	Birth of daughter Irène
1898	Pioneer research into radioactivity; begins collaboration with husband Pierre. Discovery of polonium and radium
1899–1903	Working in shed-laboratory isolating pure radium from pitchblende
1903	Presents doctoral thesis on radioactivity. Wins Nobel Prize for physics with Pierre and Becquerel for work on radioactivity
1905	Birth of second daughter Eve
1906	Pierre Curie killed in traffic accident
1910–11	Langevin scandal
1911	Wins Nobel Prize for chemistry for discovery of polonium and radium
1914–18	Assembles mobile radiography units, which are driven to the front
1916	Passes driving licence. Start of collaboration with daughter Irène
1918	Radium Institute opens in Paris

1921   Trip to America with two
        daughters
1932   Inauguration of Radium Institute in
        Warsaw
1933   Joliot-Curies discover 'artificial
        radioactivity'
1934   Dies of cancer due to radiation
        exposure at the age of 66
1935   Daughter Irène Joliot-Curie wins
        Nobel Prize for chemistry

SUGGESTIONS FOR FURTHER READING

Quinn, Susan: *Marie Curie: a Life* (Simon & Schuster, 1995) – The latest full-scale biography

Curie, Eve: *Madame Curie* (Frequent Editions) – The official saintly version of her life by her daughter: many fascinating personal details

Bernstein, Jeremy: *A Theory for Everything* (McGraw-Hill, 1996) – Range of essays on modern science, including a chapter on Curie

Romer, Alfred: *The Discovery of Radioactivity and Transmutation* (Dover, 1964 et seq)

EINSTEIN & RELATIVITY

$$e = mc^2$$

Few equations have entered our consciousness with the speed and impact of Einstein's cosmos-changing formula. From the moment in 1905 and 1917 he published his revolutionary papers on his Theory of Relativity, mankind's view of the world and the universe changed forever, the latest phase of the modern age was born, and our horizons shifted.

But how many of us really know what his theory really means, and what implications it has? *Einstein & Relativity* presents a brilliant snapshot of Einstein's life and work, together with their historical and scientific context, and gives a clear and accessible explanation of the meaning and importance of Einstein's Theory of Relativity, and the way it has changed and shaped our thinking in the twentieth century.

The Big Idea is a fascinating series of popular science books aimed at scientists and non-specialists alike. Science is at its most exciting and gripping at moments of great discovery, and each of the books in the series looks in depth at the great moments that have advanced mankind's scientific knowledge and at the men and women who have made these huge breakthroughs in our thinking about the universe and our place in it.

3 8002 00761 8053

OTHER TITLES AVAILABLE IN *THE BIG IDEA* SERIES

ALL ARROW BOOKS ARE AVAILABLE THROUGH MAIL ORDER OR FROM YOUR LOCAL BOOKSHOP AND NEWSAGENT

PLEASE SEND CHEQUE/EUROCHEQUE/POSTAL ORDER (STERLING ONLY), ACCESS, VISA, MASTERCARD, DINERS CARD, SWITCH OR AMEX

EXPIRY DATE .... SIGNATURE...............................

PLEASE ALLOW 75 PENCE PER BOOK FOR POST AND PACKING U.K.

OVERSEAS CUSTOMERS PLEASE ALLOW £1.00 PER COPY FOR POST AND PACKING.

ALL ORDERS TO:
ARROW BOOKS, BOOKS BY POST, TBS LIMITED, THE BOOK SERVICE, COLCHESTER ROAD, FRATING GREEN, COLCHESTER, ESSEX, CO7 7DW

NAME...

ADDRESS..

....

Please allow 28 days for delivery. Please tick box if you do not wish to receive any additional information ☐

Prices and availability subject to change without notice.